ICME-13 Topical Surveys

Series editor

Gabriele Kaiser, Faculty of Education, University of Hamburg, Hamburg, Germany

More information about this series at http://www.springer.com/series/14352

Irene Biza · Victor Giraldo
Reinhard Hochmuth · Azimehsadat Khakbaz
Chris Rasmussen

Research on Teaching and Learning Mathematics at the Tertiary Level

State-of-the-art and Looking Ahead

 Springer Open

Irene Biza
School of Education and Lifelong Learning
University of East Anglia
Norwich
UK

Victor Giraldo
Universidade Federal do Rio de Janeiro
Rio de Janeiro, Rio de Janeiro
Brazil

Reinhard Hochmuth
Institut für Didaktik der Mathematik
  und Physik
Universität Hannover
Hannover
Germany

Azimehsadat Khakbaz
Bu-Ali Sina University
Hamadan
Iran

Chris Rasmussen
Department of Mathematics and Statistics
San Diego State University
San Diego, CA
USA

ISSN 2366-5947   ISSN 2366-5955   (electronic)
ICME-13 Topical Surveys
ISBN 978-3-319-41813-1   ISBN 978-3-319-41814-8   (eBook)
DOI 10.1007/978-3-319-41814-8

Library of Congress Control Number: 2016943382

Printed on acid-free paper

This Springer imprint is published by Springer Nature
The registered company is Springer International Publishing AG Switzerland

Main Topics You Can Find in This ICME-13 Topical Survey

- Mathematics teaching at the tertiary level
- The role of mathematics in other disciplines taught at the tertiary level
- Textbooks, assessment and students' studying practices
- Transition to the tertiary level
- Theoretical–methodological advances.

Contents

Research on Teaching and Learning Mathematics at the Tertiary Level: State-of-the-Art and Looking Ahead

1 Introduction

Research in tertiary mathematics education is a fast growing field as evidenced by the success of the *Topic Study Group on Mathematics Education at Tertiary Level* in the *International Congress on Mathematical Education* (*ICME*); the continued success of the *Research in Undergraduate Mathematics Education conference in the United States* (RUME), now at its 19th annual conference; the university mathematics research contribution to the *Espace Mathématique Francophone* (EMF) since 2006; and since 2011 the working group on *University Mathematics Education* in the *Congress of European Research in Mathematics Education* (*CERME*) conference. In 2015 the *Australian Mathematical Society* established a *Special Interest Group in Mathematics Education*, which has among its goals the promotion of inquiry and discussion about tertiary mathematics education. Additionally, the first conference of the *International Network for Didactic Research in University Mathematics* (*INDRUM*) launched in Montpellier, France in March 2016. Furthermore, in 2015 the new *International Journal of Research in Undergraduate Mathematics Education* published its first issue and as we write these lines the following three handbook chapters will appear: *Post-calculus research in undergraduate mathematics education* (Rasmussen and Wawro in press), *Understanding the concepts of calculus: Frameworks and roadmaps emerging from educational research* (Larsen et al. in press) and *Tertiary mathematics education research in Australasia* (Coupland et al. 2016, in press).

The literature review we offer here contributes to the three aforementioned handbook chapters with additional literature, mainly from research outcomes published since 2014. Moreover, to not repeat and to emphasize the most recent advances not captured in the handbook chapters we include results disseminated in conferences since 2014 as indicators of the current trends of the field. We were judicious in our choices of which conference papers to include, using only those that, in our view, were of high quality.

© The Author(s) 2016
I. Biza et al., *Research on Teaching and Learning Mathematics at the Tertiary Level*, ICME-13 Topical Surveys,
DOI 10.1007/978-3-319-41814-8_1

2   Survey and State-of-the-Art

The five emergent areas of interest we identified and reviewed are mathematics teaching at the tertiary level; the role of mathematics in other disciplines; textbooks, assessment and students' studying practices; transition to the tertiary level; and theoretical-methodological advances. We conclude with a summary on the literature we reviewed and with suggestions for ways forward. Due to the space limitations we have included publications related to student learning of specific mathematical topics (e.g., Calculus, Linear Algebra, etc.), proof and reasoning, statistics education and mathematics teacher education, which are more relevant to other topic study groups at ICME13, only if their input goes beyond these topics and contributes to our understanding of one of the five emergent areas of interest.

2.1   Mathematics Teaching at the Tertiary Level

Six years after the observation of Speer et al. (2010) on the existence of very little empirical research on actual teaching practices at university level, we are now in the pleasant position to report ongoing growth and interest in this area. This section explores new work on tertiary mathematics teaching, focusing on the themes of: mathematics teaching practices; the influence of teachers' perspectives, background and research practices on their mathematics teaching; resources and preparation for teaching and teachers' professional development; and alternative approaches to teaching. In this section teaching takes place at the tertiary level. The domains of teaching at secondary level and secondary school teacher preparation, which is under the prevue of different ICME13 Topic Study Groups, is beyond the scope of this survey.

2.1.1   Mathematics Teaching Practices at the Tertiary Level

In this section we discuss mathematics teaching practices at the tertiary level with specific emphasis on three recent foci: communication of mathematical ideas, mathematics teaching activity in relation to its goals and intentions, and characteristics of university mathematics teaching.

Communication of mathematical ideas has been the focus of several recent studies that consider mathematical teaching as a communication act with attention on mathematical and pedagogical discourses, including gestures. From this perspective, Viirman (2014, 2015) sees university mathematics teaching as a highly situated discursive practice. Through the lens of Sfard's Commognitive framework (2008) he studied the discursive teaching practices of seven mathematics teachers on the topic of functions. In Viirman (2014) he focuses on the mathematical discourse and offers a categorization of the construction and substantiation routines used by the teachers. Later, in Viirman (2015), a more refined analysis indicates that, although teachers' mathematical discourse has similarities to each other regarding the word use and visual mediators, their didactical discourse has different routines. The routines of participants' didactical discourse were categorized as *explanation, motivation,* and *question posing* routines. Within these general categories, a number of different sub-categories of routines were found, each used in different ways and to different extent by different teachers. According to Viirman (2015) this classification of routines, although not meant to be complete and developed in only a single content context, can also be useful in the investigation of teaching other mathematical topics.

A discursive approach was also used by Park (2015) in the analysis of the teaching of derivative in calculus courses. The study analyzes features of three teachers' classroom discourse on the derivative in relation to word uses, gestures and the use of different visual mediators (symbolic, algebraic, and graphical), emphasizing the connection between graphical and symbolic representations, and the transitions between a point-specific and an interval view of the derivative. The results indicate that the use of words and visuals were consistently reflected in teachers' routines and endorsed narratives. Moreover, these three teachers seem to assume that a point-specific and an interval view of the derivative are clear to their students without making explicit reference to them, although they were in their endorsed narratives. This indicates a disconnection "between the endorsed narrative of the teacher and the students' abilities to comprehend what the teacher is saying" (ibid p. 248).

Continuing with how university mathematics teachers communicate mathematical ideas, Weinberg et al. (2015) describe teachers' gestures and how these contribute to opportunities to communicate mathematical ideas in an abstract algebra lecture. Gestures, and specifically teacher use of the hand and fingers to point during undergraduate mathematics lectures, is the focus of a study from Hare and Sinclair (2015), who observed a teacher in a third-year Group Theory lecture. Results suggest that pointing seems to act as a significant interface between speech and written text on the board. The authors argue that gestures are important components of communication as they bring "mathematical objects into being," relate "these objects to each other," and connect "the spoken with the written and drawn" (p. 33).

Other studies consider university mathematics teaching as a goal-oriented activity and adopt an Activity Theory (Leontiev 1978) lens to identify the connections between a teacher's goals and actions. Grenier-Boley (2014) suggests a methodological approach for the investigation of how linear algebra concepts with formalizing, unifying, and generalizing characteristics such as vector spaces are introduced during tutorial sessions at the beginning of university studies. Teaching as a goal-oriented activity is also the focus of Treffert-Thomas (2015) in her theorization of linear algebra teaching to first year mathematics undergraduates. Through the analysis of lecture observations and interviews with one teacher, Treffert-Thomas categorized the teaching in three levels: *activity-motive, actions-goals* and *operations-conditions.* Furthermore the study suggests an action-goal model of the teaching process that relates lecturer's intentions (as expressed in the interviews) and the strategies (the actions and goals) that he designs for his teaching.

Some recent studies focus on the role of exemplification in mathematics teaching in lectures (e.g., Fukawa-Connelly and Newton 2014) or in small group tutorials (Mali et al. 2014). Furthermore, Mali et al. (2014) suggest the use of examples as one of the characteristics of mathematics teaching, where a characteristic refers to patterns in the ways that tutors teach in the tutorials, such as the use of representations and problem solving techniques. Teaching characteristics, operationalized as actions and rationale, are also the aim of Petropoulou et al. (2015), who investigate the mathematics teaching practices of one teacher in calculus lectures. The analysis indicates how the teacher shows sensitivity to students' needs and draws them into the mathematical culture by offering mathematical challenges.

2.1.2   The Influence of Teachers' Perspectives, Background, and Research Practices on Their Teaching

Recent studies suggest that lecturers' research practices influence their approaches to teaching. For example, Mali et al. (2014) discuss the use of generic examples in small group tutorial teaching and suggest that the research practices of the tutor, in this case a research mathematician, influences her teaching practices. Similarly, Mali (2015) analyzes two teaching episodes from two tutors: one with a doctorate in mathematics and one with a doctorate in mathematics education. The two tutors follow different approaches even for the same didactical choice (e.g., the use of graphical representations). This study raises the issue of how and to what extent teachers' research background and their potential epistemological position influence their teaching practices. These observations accord with similar finding by Petropoulouet al. (2011) about teaching practices in lectures.

Additionally and from a different perspective, research indicates that beliefs about teaching and learning of mathematics graduate teaching assistants, who play significant role in undergraduate mathematics teaching in the US (Ellis 2014), have a significant impact on their classroom practices and decisions (Kim 2011).

2.1.3   Resources and Preparation for Teaching—Mathematics Teachers' Professional Development

The focus here is mathematics teachers' activities *outside the class*, especially in relation to teaching preparation and professional development (PD). Gueudet and colleagues investigate university teachers' practices in their preparation and delivery of teaching, as well as their communication with each other and their students. These studies adopt the documentational approach (Gueudet et al. 2012), considering teachers' interaction with resources and how this influences their knowledge and practices (see Gueudet et al. 2014 on this approach in the context of university mathematics). Gueudet (2015), for example, utilizes the documentational approach to analyze interviews with six teachers and their interaction with resources in a goal-oriented activity that produces documents and documentation systems, namely structured set of all the documents they develop.

Recently, there is an increasing interest by tertiary teachers in non-lecture pedagogies. This interest has been reflected in the research in this area. For example, Hayward et al. (2015) report on the impact a series of annual, weeklong PD workshops for college mathematics teachers on Inquiry-Based Learning (IBL) in undergraduate mathematics had on their teaching practice. 58 % of the teachers reported implementing IBL strategies in the year following the workshop they attended.

Additionally, there is an emerging need for PD provision available for graduate students who contribute to teaching. Rasmussen et al. (2014) studied five exemplary calculus programs at US institutions that offer a doctoral degree in mathematics in order to identify the key common characteristics. Among the seven programmatic and structural features common to these institutions was graduate teaching assistant training, with the other six being: coordination; attending to local data; active learning; rigorous courses; learning centres; and placement.

2.1.4 Alternative Approaches to Teaching

There is a range of alternative approaches of teaching in the literature we reviewed, especially towards a reduction of the traditional lecturing model. Oh Nam (2015) describes an implementation of *flipped learning* approach where traditional lectures and homework are reversed and students attend the short instructions in videos or online courses while face-to-face time is devoted to classes for exercises, activities, or discussions. Alternatively, in *peer assisted learning* (PAL), students assist other students in the development of study skills. Duah et al. (2014) adopted a PAL approach to assist students in their transition to university mathematics studies and the reduction of the *cooling off* phenomenon (i.e., when undergraduate mathematics students lose their motivation and interest for their studies).

Another mathematical learning opportunity for undergraduate students is their involvement in curricular and teaching/learning resource development. Although there is little research in this area, there is a growing interest in students' engagement as partners in course design. Croft et al. (2013) identified several benefits for students from their involvement in producing screencasts for other students, such as better and deeper understanding of mathematical topics, development of technological skills, improvement of study habits, personal and organizational skills and enhanced communication skills. Solomon et al. (2014) see undergraduate students' internships as a pathway for improving dialogue between students and staff that challenges traditional hierarchical roles and relationships. Similarly, Biza and Vande Hey (2014) in their evaluation of students' learning through their involvement in statistics teaching resource development projects, deployed the *Communities of Practice* approach (Wenger 1998) in order to consider not only the learning per-se, but also how and in what extent this learning is a result of students' participation. This evaluation revealed that students explicitly linked their participation to the solidification and organization of their knowledge in terms of statistical thinking and reasoning. They also found out how students merged, or sometimes experienced conflict with, multiple perspectives—such as student versus developer and mathematician versus non-mathematician (Biza and Vande Hey 2014).

2.2 The Role of Mathematics in Other Disciplines Taught at the Tertiary Level

In this section we discuss the role of mathematics in other disciplines taught at the tertiary level, such as engineering, science, economics, etc. Although there is a substantial body of research in teaching mathematics in teacher education programs (see for example, international research projects such as TEDS-M, Blömeke et al. 2014), this area is under the prevue of other ICME-13 Topic Study Groups and, thus will not be addressed here.

The increased interest in research on teaching mathematics within study programs of other disciplines is compatible with the general growing interest on didactical issues at the tertiary level, but it is also connected to the relevance of mathematics in other disciplines and in particular the growing importance of quantitative methods in general (National Research Council 2003, 2009). Further relevance arises from the large number of students enrolled in those study programs and the high failing and dropout rates (Heublein et al. 2014; Ulriksen et al. 2010). Although mathematics lectures in these programs are often different from those for mathematics majors, the students experience similarly a higher level of mathematics as compared to school (Mustoe 2002). Harris et al. (2015) investigated the influence of students' mathematics experiences on their disposition towards their engineering courses and mathematics. Their longitudinal study identified as one major problem that the students need to get insight of the real use of mathematics in their discipline.

In several countries, government-funded programs intend to enforce study reforms for improving the situation. For example, in England, Tolley and MacKenzie (2015) give an overview on support provided to non-mathematics majors though Mathematics Learning Support Centres (MLCS) in higher education, report on research undertaken to establish a management perspective on the provision of mathematics support and figure out expected future needs and scope for national collaboration and coordination. Goodchild and Rønning (2014) report about a large research and development project at the Norwegian University of Science and Technology (NTNU) intended to modernize the mathematics education of engineering students. In Germany bridging courses of different types and with various foci are the most prominent method for supporting students in their transition from school to university, see for example (Bausch et al. 2014).

The following paragraphs summarize recent research on teaching mathematics within other study programs. First we report on studies about major challenges in this field and proposals for enhancing teaching. Then, we discuss the relation between mathematics and its application, in particular modelling. We conclude with remarks on specific mathematical topics in non-mathematics programs.

2.2.1 Teaching Mathematics to Non-mathematics Students: Challenges and Proposals for Enhancing Teaching

Harris et al. (2015) analyze the problems engineering students have with mathematics and lecturers on the basis of interviews. The authors challenge both the pedagogical practice of teaching non-contextualized mathematics and the lack of transparency regarding the significance of mathematics to engineering and "conclude that the value of mathematics in engineering remains a central problem" (p. 321). A common practice for enhancing teaching is to complement traditional courses by offering video recorded lectures. Rønning (2016) reports from surveys at NTNU since 2013 showing students' preference for traditional courses, which could be traced, on the basis of interview data, to the provision of structure in students' daily life, a sense of good conscience visiting the lectures, and opportunities to communicate to fellows and lectures.

Fostering conceptual understanding of mathematics by first year engineering students is the reported goal of several teaching modifications. Albano and Pierri (2014) report about a role-play activity and a competencies-based analysis, based on the competency frameworks of Niss (2003) and by the European Society for Engineering Education (Alpers et al. 2013). The basic idea was that students create materials themselves. To this end mathematical content has been split into suitable parts and for each part students create a cycle of activities based on role-play. The qualitative analysis suggests a move from an instrumental towards to a more relational approach in students' learning. Loch and Lamborn (2016) report about a project where advanced engineering students and multimedia students worked on two animated videos demonstrating the application of first year mathematics by professional engineers. Interviews with advanced students and focus groups with first year students indicate differences in their views how to make such interviews. The authors suggest that such videos should be produced by collaborative groups of advanced and first year students, advised by lecturers.

Jaworski and Matthews (2011) report on a project that is related to a mathematics module for materials engineering students in a UK university with the goal to achieve a more conceptual understanding by students. Past modifications had shown little effect, but a new more coherent and far-reaching innovation including inquiry based small group activities, "a variety of forms of questioning, an assessed group project and use of the GeoGebra-medium for exploring functions" (p. 178), changed the situation. The analysis focuses on the students' participation in the teaching innovation. It is based on the community of practice (Wenger 1998) and documentational approaches (Gueudet et al. 2014). Although there was evidence of students' understanding, other aspects of their studies and especially the assessment by examination distracted students from their alignment to such initiative's objectives (Jaworski et al. 2012).

Following the Theory of Didactic Situations (Brousseau 1996) and the Anthropological Theory of Didactics (Chevallard 2003), Barquero et al. (2008) described and analyzed the implementation of so called "study and research groups" in a one-year "Mathematics Foundation of Engineering Course" of a technical engineering study program and Serrano et al. (2010) in a mathematics course of a first year Business Administration and Management study program. Barquero and Bosch (2015) discussed these teaching innovations regarding their epistemological dimension with a specific focus on the didactical engineering.

One of the few papers giving empirical evidence that teaching innovations lead to lower failing and dropout rates is by Hieb et al. (2015), where the impact of an algebra readiness exam was analyzed. The previous result that the readiness exam scores are a significant predictor of retention and performance in an Engineering Analysis I course was verified. Moreover, a hierarchical linear regression model was created showing that time and study environment management, internal goal orientation, and test anxiety significantly predict exam scores.

2.2.2 Mathematics and Its Application, Particularly in Modelling

There are only a few studies investigating the specific meaning of mathematical concepts within applications. Prominent models, at least in school, for describing the application of mathematics to real world problems are modelling cycles. Therefore it seems natural to apply this tool also in the tertiary mathematics context. Czocher (2014) question in a task-based interview study whether this really makes sense. Her qualitative study shows that the mathematical thinking involved in mathematical model construction is neither sequential nor quasi-periodic, confirming prior conjectures that "the view presented on modelling as a cyclic process is highly idealized, artificial, and simplified" (p. 359). This is in line with results by Hochmuth et al. (2014) considering mathematical practices in signal analysis and also by Biehler et al. (2015). The latter paper discusses the development of mathematical skills required in technical subjects of engineering bachelor courses and focuses on a foundation of electrical engineering courses offered at a German university. The study analyzes task-based interviews with experts and draws on the literature on modelling (Blum and Leiß 2007; Kaiser and Brand 2015), problem solving (Polya 1949), and mathematical argumentation and resources in physics (Bing 2008; Tuminaro and Redish 2007). From their analysis they tentatively suggest that it is counterproductive to try to separate the mathematical and "real world" (engineering) parts of the problem. Concerns about the usefulness of the modelling cycle were recently also articulated in the physics education context by Uhden et al. (2012).

Frejd and Bergsten (2016) analyze mathematical modelling as a professional task. They investigate scholarly knowledge in an interview study with nine professional mathematical model constructors from the perspective of didactic transposition processes. The grounded theory inspired analysis of the interview data led to categorization of three main types of modelling activities: data-generated modelling, theory-generated modelling, and model-generated modelling. The use of computer support and communication between clients, constructors and other experts turn out to be always central aspects.

In another interview study with two professional engineers with very different backgrounds by Bergsten et al. (2015) it turned out that a conceptual mathematical approach was highly relevant to the engineering work in which the interviewees were involved, although "these engineers did not perform much mathematical activities in terms of actual calculations" (p. 988). Moreover the importance of "a more general engineering understanding" and of independent work, self-confidence, and judgment became thematic (p. 988). Engelbrecht et al. (2015) confirmed the emphasis on conceptual approaches through interviews with 23 other engineers.

2.2.3 Mathematical Concepts in Non-mathematics Programs

Alpers (2011) reported about a four-year workplace study identifying the mathematical expertise a mechanical engineer needs. The author focuses in particular on competences "regarding the effective and efficient use of computational tools and qualitative models in mechanical engineering tasks" (p. 2), which is rather relevant for teaching innovations. Hochmuth and Schreiber (2015) investigated the mathematical discourse in Signal and System Theory courses (SST) applying a praxeological model based on concepts from the Anthropological Theory of Didactics (Chevallard 2003), which allows in particular discriminating between practices and reasoning patterns established in higher mathematics courses and in SST-courses.

Hester et al. (2014) noticed that biology students are often not able to apply quantitative skills in biological contexts. Therefore an introductory molecular and cell biology course was established in which the authors integrated the application of mathematical skills with biology content throughout. A pre/post course outcome assessment shows that the students are more successful in integrating mathematics and biology items and "made comparable gains on biology items, indicating that integrating quantitative skills into an introductory biology course does not have a deleterious effect on students' biology learning" (p. 54).

Ramful and Narod (2014) investigated ways in which proportional reasoning is involved in the solution of chemistry problems in stoichiometry. The main result shows, that the proportionality tasks arising in chemistry are more complex and difficult as compared to those in the mathematics curriculum. Jukić Matić and Dahl (2014) report about a case study considering the retention of differential and integral calculus concepts of a second-year student of physical chemistry at a Danish university. Their main result is that a successful application of calculus in a physical chemistry study program does not strengthen the original calculus learnt.

Mathematics also plays an increasing role in business administration and economics programs. Mkhatshwa and Doerr (2015) investigate economics students' reasoning about topics from calculus (instantaneous rate of change) in an economic context (marginal change). The analysis indicates that the distinction between amount and rate of change is rather difficult for students who deal with these concepts in non-mathematical contexts. Ariza et al. (2015) linked students' understanding of the function-derivative relationship "to students' capacity to perform conversions and treatments between the algebraic and graphic registers of the function-derivative relationship when extracting the economic meaning of concavity/convexity in graphs of functions using the second derivative" (p. 615).

Mills (2015) applies an online survey instrument to figure out "what calculus topics business faculty view as relevant to, and necessary for, various business specializations" (p. 231). The most frequently mentioned concepts were differentiation, optimization, and rate of change. Instructors from Accounting, Entrepreneurship, and Management departments did not list any calculus concepts. Economics and Finance professors state "that integration was not used in their undergraduate courses, and that the masters and doctoral students who would use integration were required to have a more rigorous mathematical background (up through differential equations) before entering the program" (p. 235).

2.3   Textbooks, Assessment and Students' Studying Practices

At the tertiary level students are expected to spend a considerable amount of time out of class studying course material (e.g., the textbook, notes, other resources) and working homework problems. Indeed, in courses that are primarily lecture oriented, students' out of class practices often constitute the primary set of learning opportunities. In this section we review the literature on what we know about the opportunities textbooks typically afford learners and how students use and read textbooks and online resources. Finally, we discuss students' role in assessment practices.

2.3.1   Opportunities Afforded by Textbooks

Certainly what students might get out of a text is a function of what opportunities the textbook affords. Raman (2002) uses two popular textbooks in the US, one precalculus textbook and one calculus textbook, to illustrate ways that these texts likely impede occasions for students to make connections between informal and formal aspects of mathematics. Using continuity as a case in point, Raman argues that these texts tend either to emphasize the formal at the expense of the informal or vice versa. More specifically, Raman found that on the one hand, the precalculus text does not offer opportunities for students to use the informal characterization of continuity that is provided in the text. On the other hand, the calculus text provides the formal definition of continuity but then does not offer opportunities for students to construe informal or intuitive meanings for the definition. Moreover, neither text offers opportunities for students to compare an informal meaning of continuity with the formal definition. Raman similarly found that the two texts treat graphs and theorems in ways that offer little opportunity for students to reconcile their informal or intuitive ideas with the more formal mathematics. In a related analysis, Raman (2004) carries out an epistemological analysis of these same two texts together with a popular real analysis textbook, finding that these texts send different, and at times conflicting, messages about the status and role of mathematical definitions. The practical implications of Raman's insights are clear: textbook authors would do well to consider more carefully how exposition and problem sets can offer opportunities to relate informal and formal mathematical ideas.

In a different context, González-Martín et al. (2011) analyzed 22 texts used in Canada and UK post-compulsory courses to identify how the concept of infinite series is introduced and found that the presentation of the concept is largely a-historical, with few graphical representations, few opportunities to work across different registers (algebraic, graphical, and verbal), few applications, and few conceptually-driven tasks.

Lithner (2004) also examined the opportunities that a common Swedish calculus textbook affords students. In particular, using a reasoning framework developed in an earlier study (Lithner 2000), he investigated the possible ways for students to solve problems without considering the intrinsic mathematical properties involved and how frequently such problems appear. Lithner found that approximately 90 % of the problems could be solved by searching for similar worked out problems (similar in the sense of surface properties), by mimicking the solution procedure, or by what he refers to as local plausible reasoning. Local plausible reasoning is similar to mimicking but the student must determine if the solution procedure can be copied or determine how to make a slight modification to the solution. What students are asked to do on homework is of course reflected in what is valued in class and on examinations. Tallman et al. (in press) examined 150 Calculus I final examinations in the US and what they found is compatible with the analysis conducted by Lithner (2004). Approximately 85 % of problems required only rote knowledge or simply recall and application of a procedure.

Mesa (2010) took analysis of calculus textbooks a step further and examined not only the afforded solution strategies but also the opportunities embedded in a text for students to verify that their solution is correct. To accomplish this goal Mesa analyzed 80 examples of initial value problems (IVPs) across 12 different calculus textbooks. IVPs were selected because they offer opportunities to bring together integration, differentiation, and real-life situations and hence the subject-milieu system (Balacheff and Gaudin 2010) potentially offers more opportunities to verify solutions. Such opportunities, however, were not realized. Mesa found that of the 80 examples analyzed, only about a third of them demonstrated how to determine if the answer was correct or appropriate.

From a different perspective, Park (2016) used a Commognitive approach (Sfard 2008) to analyze word use and visual mediators in the presentation of the limit process in which the derivative as a function is objectified in three textbooks in the US. The analysis identified inconsistencies, implicit transformations, and disconnections between visual mediators and objects. Also, both the derivative at a point and the derivative of a function were mediated with nearly identical symbols, potentially constraining the students' understanding of the differences between them.

In contrast to these studies that point to the limitations of textbook problems and final examinations, White and Mesa (2014) found that the assigned homework problems at a particular community college were more cognitively demanding than one might expect in light of previous research. Using a similar task analysis framework to that used by Tallman et al. (in press), White and Mesa found that nearly half of the assigned tasks were classified as complex procedures or rich tasks. This is strikingly higher than that found by Tallman et al. in their study of 150 final exams. One reason for this higher than expected level of cognitive demand is likely related to the fact that this particular Calculus I program investigated was identified in a US national study as having a more successful Calculus I program, where success was measured by pass rate, persistence rate to Calculus II, and changes in student confidence, interest, and enjoyment of mathematics (Bressoud et al. 2015). Similar findings about the challenging nature of the assigned textbook problems were also identified in five doctoral degree granting institutions identified as having a successful Calculus I program (Ellis et al. 2015; Bressoud and Rasmussen 2015). Challenging textbook assignments alone, however, cannot account for the success of particular Calculus I programs. As detailed by Bressoud and Rasmussen (2015), challenging assignments was just one of seven characteristics that capture the system level organization of calculus instruction. However, it is noteworthy that the more successful Calculus I programs increased the cognitive demand of assignments and that students were supported to rise to the challenge.

2.3.2 How Students Use and Read Textbooks

In the previous section we reviewed the research that examined the opportunities that textbooks afford students. The overall finding from these studies is that textbooks in introductory courses often fall short of providing an abundance of rich, cognitively demanding problems. Based on reader-oriented theory, Weinberg and Wiesner (2011) underscore the importance of attending to how students actually use texts, the beliefs of the reader, and the affordances and constraints of the text itself. With these features in mind, we next provide a brief review of studies that examine how students use and read textbooks.

Lithner (2003) video-recorded three Swedish students individually working through calculus homework problems in their usual manner for two hours, followed by a post-interview. Two students were computer engineering majors, one with high grades and one struggling in the course. The third student was a prospective upper secondary school mathematics teacher, also struggling in the course. Despite these differences in students, Lithner found that almost all of their time was spent on searching for surface similarities with a procedure or worked out example in the textbook. Compatible with the analysis of IVPs by Mesa (2010), Lithner also found that students engaged in very little justification or verification of their solution—following the textbook's example was authority enough. In terms of what students likely learned from their homework, Lithner found little to no evidence that these students attempted to learn more global properties or general ideas associated with the problems.

In a much larger scale study, Weinberg et al. (2012) surveyed 1156 undergraduates at three different institutions in the US about how they use their mathematics textbook. Similar to the in depth study of the three students by Lithner (2004), Weinberg et al. (2012) found that students use worked out examples to guide their solutions. Reading the expository text for meaning and more general ideas was not a common strategy. Consistent with previous studies, students mainly reported using exercise solutions to verify the correctness of their solutions. Weinberg et al. (2012) also surveyed students on how frequently their instructor asked them to read the exposition in their text. Interestingly, even for students in the same classroom, there was significant variation in how frequently they reported their instructor requesting them to read the textbook. Nonetheless, when students reported that their instructor asked them to read their text they are more likely to do so. This suggests that instructors can play a role in increasing the extent to which students read the exposition in their textbook, which is typically where authors attempt to convey the big ideas and underlying meanings.

It seems clear that university students tend to use textbooks in ways that are unintended by the authors. Instead of reading the exposition for meaning and big ideas, they tend to search for similar worked out problems and mimic or make small adjustments to the example procedures. However, even when students do read the text, Shepherd et al. (2012) found they are not effective in reading mathematics. This finding was based on analysis of 11 precalculus and calculus students who all had high ACT[1] mathematics, high ACT reading comprehension scores, and who exhibited many of the behaviors of good readers as described in the reading comprehension research literature. Approximately halfway through the course, each student was given a passage from their text to read, asked to think aloud, and then invited to solve a few straightforward tasks directly related to the reading. Even though these students were good readers and were mathematically competent, all had difficulty completing the straightforward tasks. These authors were not able to trace student difficulties to the writing style of the textbooks. Instead, Shepherd et al. argue that reading mathematical text appears to call on different abilities from those required by the two ACT tests.

The studies reviewed here point to the at-times ineffective reading strategies of relative novices, but how do these strategies compare to those of experts?

---

[1]In the US, the ACT is a national college admissions examination.

Knowledge about these differences has the potential to inform training on reading strategies for beginning undergraduate students. To begin to address this line of inquiry, Shepherd and van de Sande (2014) compared three mathematics faculty members' and three mathematics graduate students' readings of an unfamiliar graduate textbook to the reading strategies of first-year undergraduate students reported in Shepherd et al. (2012). They identified three overarching differences: First, the graduate students and mathematics faculty members were more likely to read the meaning of symbols rather than reading the symbols verbatim. Second, the graduate students and faculty exhibited more meta-cognitive awareness of what they did and did not understand. For example, they spent more time working through ideas and more frequently checked for understanding. Third, the more advanced readers explored the content they read by creating examples and referring to supplementary representations when needed. For example, the undergraduates tended to read lengthier passages without pausing to check for understanding whereas the more advanced readers stopped nearly twice as often.

The findings about how experts and novices read texts lead to questions about how experts and novices read proofs. The research in this area largely point to similar differences between experts and novices. For example, compared to novices, mathematicians tend to be more metacognitively aware of their comprehension and employ a wider range of strategies to foster comprehension (Weber 2015; Weber and Mejia-Ramos 2011; Yang 2012). In one of the few studies that has designed and investigated the effect on a specific intervention on students' comprehension of proofs, Hodds et al. (2014) found that self-explanation training increases student proof comprehensive. The intervention was a fairly straightforward booklet that focuses on logical relationships. We see this study as an important step forward in that it builds on prior research to design and implement an effective strategy for improving student reading and comprehension. The field is in need of fewer deficit accounts and more interventionist studies that seek to improve student learning.

2.3.3 How Students Use Online Resources

One of the findings by Shepherd and van de Sande (2014) was that experts read text differently than novices—they stop more often to check for understanding. This finding raises an interesting question related to the use of videos. One of the features of instructional videos is that they allow students to pause, rewind, and rewatch whenever they chose. This feature potentially allows beginning undergraduates to more easily stop and check for understanding. So does this feature bring undergraduates into closer contact with the reading practices of experts? The analysis by Le et al. (2010) suggests that the answer is no. Le et al. (2010) investigated a blended learning approach at a Canadian university where students could attend the lecture live and/or watch a video recording of the lecture. Previous studies in an introductory psychology course showed that student use of the pause and seek feature was positively correlated with exam grades (Bassili and Joordens 2008).

In contrast to these findings, Le et al. found students who both attended lecture and watched online and those who made the most use of the pause and seek options were precisely those that were the poorest performers in the course. This finding was consistent across student ability range, and thus could not be attributed to differences in general mathematics ability. As an alternative hypothesis, Le et al. conjectured that the negative correlation between exam performance and use of the online media features reflects differences in learning strategies. To test this hypothesis they repeated the study in a second semester calculus course. They found that those students who primarily seek to memorize material used the pause feature more often and performed worse in the course. This finding points to the need to help students develop more effective strategies for using and learning from online content.

A potential shortcoming in the Le et al. studies is that the analysis relied on student self report of their use of online media. To in part address this shortcoming, Inglis et al. (2011) investigated actual student use of resources in multivariable calculus course in England. More specifically, these researchers tracked the student usage of online lectures, attendance at face-to face-lectures, and visits to the mathematics learning support centre. Consistent with Le et al., Inglis et al. also found that those students who most often watched online lectures had lower grades that those students that attended lecture or made use of the support centre.

2.3.4 Students and Assessment Practices

Iannone and Simpson (2015a, b, in press) investigated the ways in which undergraduate mathematics students perceive summative assessment. They found that students' perceptions differ significantly from what is reported by the general educational literature. Indeed these students prefer to be assessed by assessment methods they perceive to be good discriminators of ability and see the closed book examination as one of the best of these discriminators (Iannone and Simpson 2015a). They also see oral examinations as a very good way to assess for ability in mathematics. In a follow up study (Iannone and Simpson 2015b) show how English students appreciate the oral examination method for its immediacy of feedback and perceive this assessment method as requiring conceptual understanding. Iannone and Simpson (in press) also investigated the reasons why mathematics students' perceptions are different from those reported in the literature via a comparative study between mathematics and education students and found very different results. It would appear that students' discipline grounded epistemic beliefs are a very big component in shaping their preferences and those beliefs do vary significantly across disciplines.

Jones and Alcock (2013) tested an alternative approach of peer-assessment where students judge peers' work against assessment criteria with pairwise comparative judgment where students judged pairs of scripts against one another in the absence of assessment criteria. 194 first year mathematics undergraduates took a written test on multivariable calculus and then they assessed their peers' responses. Validity was investigated by correlating peer assessment outcomes with assessments by experts and novices as well as with marks from other module tests. High validity results suggest that the students performed well as peer assessors.

2.4   Transition to the Tertiary Level

Difficulties associated with the transition from school to tertiary mathematics teaching and learning has been a focus of research in mathematics education since early 1980s mostly by trying to understand the cognitive structures and processes that determine the difficult changes and restructurings that this transition involves (e.g. Tall 1992). As Artigue et al. (2007) highlight, the integration of anthropological and sociocultural approaches, especially from the mid 1990s, has broadened research orientation, which started to take into consideration the role social, cultural, and institutional practices play in the transition from school to university contexts (see also, Gueudet 2008).

In this section, we briefly review some recent research results that focus specifically on this transition. We start with addressing (dis)connections between mathematics and between students' attitudes, practices, and performances at school and tertiary level. Then, we consider the transition to abstraction and formal mathematical thinking. We close the section discussing initiatives aiming to ease the way into tertiary studies through the first year.

2.4.1   Mathematical Content in School and Tertiary Level

Recent research addresses difficulties experienced by students when starting studies at a tertiary level. These difficulties are related to differences and possibly conflicts between school and tertiary contexts. This includes teaching styles, instructional approaches, studying and learning strategies as well as views about mathematics, specific mathematical concepts, mathematical knowledge, and goals of learning.

Some studies suggest that beginning undergraduate students do not see university mathematics topics as continuations, extensions, or generalizations of topics previously studied at school—they tend to regard these as completely different subjects. For example, Cofer (2015) reports that even prospective secondary mathematics teachers were unable to link school algebra and university algebra, despite their having already finished an abstract algebra course. Such results on students' conceptions indicate a need for research focusing on how and to what extent connections between school and tertiary mathematics are established in undergraduate instruction, including teaching, syllabuses, and textbooks.

Suominen (2015) analyzed nine undergraduate abstract algebra textbooks, searching for explicit connections between abstract algebra and secondary school mathematics concepts. Results were organized according to an analytic framework based on categories adapted from previously established work by Businskas (2008) and Singletary (2012): alternate representations; comparison through common features; generalization; hierarchical inclusion; and real world applications. The author found that comparison of common features, hierarchical inclusion, and alternate representations appear more frequently than the other categories, including generalization. These results differ from earlier literature, which highlights connections related to abstract algebra as *generalizations* from school concepts (e.g., Usiskin 1974). That is, connections are established in these textbooks in a different manner. The author stresses that, as far as the connections established in the textbooks are regarded, abstract algebra "can no longer be considered simply as the generalization of school algebra but rather it should be regarded as an extension of previous mathematical knowledge from algebra and geometry" (Suominen 2015, p. 249).

2.4.2 Students' Attitudes, Practices and Performances in Secondary and Tertiary Level

Some studies try to map out possible reasons for students' difficulties in the beginning of the university from their school background, including knowledge and attitudes towards learning.

For instance, Wilkie and Tan (2015) explore school leaders' perspectives on their school-based approaches for influencing student subject choice in Australia. The study is motivated by a reported decline in the proportion of Australian students opting to study higher-level mathematics. Dweck's (2007) mindsets framework was used to investigate leaders' school-based approaches for influencing student subject choices and how these might relate to their beliefs about mathematics teaching and learning. Data were collected through audio-recorded in-depth interviews of the mathematics leaders. Schools employ a variety of strategies for influencing and guiding students' study choices in the senior years and include input from multiple sources. These strategies were often described by the leaders in terms of encouraging students to aim high and challenge themselves (growth mindset), or discouraging students from attempting a subject in which they were not deemed capable of succeeding (fixed mindset). Nonetheless, a majority of leaders indicated that students could override the recommendations and choose for themselves. A tension between the need to consider performance and the desire to promote progress was noticeable. There was also the sense that the leaders experienced tension related to differences between them and the career staff, between them and their mathematics teaching staff, or between them and parents.

Barnett et al. (2014) discuss relations between students' effort in advanced mathematics courses in US high schools and their performance in college calculus. The authors found no consistent association, and distinguish a productive effort, that carries the expected benefits, from ineffective efforts, that is associated with negative consequences. Data includes responses from a questionnaire administered to a random sample of 10,437 students enrolled in 336 college calculus courses/sections at 134 US institutions; grades in high school mathematics and college calculus; and information on daily study time and on time spent reading the course text during their most advanced high school mathematics course. The authors found that more reading of the course textbook was associated with worse college calculus performance for students with all kinds of high school preparation. "Time spent studying, however, was found to be a productive effort for students who took calculus in high school, regardless of their performance, and for high-performers who did not take calculus" (p. 1015).

Ufer (2015) investigates the relationship between students' cognitive and motivational learning prerequisites, their learning behavior and their success in the first semester of mathematics study at the university. More precisely, the paper addresses: (1) Do students' self-reported study choice motives relate to their choice of a more or less application-oriented mathematics program? (2) How do students' affective and cognitive learning prerequisites relate to their learning activities? (3) How do students' learning prerequisites and learning activities relate to their study success? Data comes from a longitudinal survey study with 333 first semester mathematics students enrolled in a regular mathematics program or a financial mathematics program at the University of Munich, Germany. Results suggest that students' motives for choosing a university mathematics program are connected to their learning behavior and study success. Students' self-reported study choice motives provided specific predictive information about their study behavior and were partially predictive for study success. In particular, clearly extrinsic motives (professional perspectives) went along with less constructive learning activities. A direct connection between a motive to apply mathematics and reduced study success was found. These results underpin the view that only motives in line with the program under study can support in coping with the transition from school to university mathematics.

2.4.3 Transition to Abstract Mathematical Concepts and to Formal Mathematical Thinking

Kempen and Biehler (2014) investigate argumentation skills of the students at the beginning of the course, and identify common gaps or pitfalls in their argumentations. The participants were 177 undergraduate pre-service teachers in Germany attending a bridging course. They were asked to verify a statement of elementary number theory, and were given a questionnaire with items concerning argumentation and proving, attitudes towards proving, the nature of mathematics, and the nature of mathematics teaching. Results reveal that participants are not equipped with the argumentation skills required for proving. This suggests that mathematics at school does not provide the future students with adequate heuristics for problem-solving and basic proving skills. These findings underline the importance of introductory courses, such as bridging courses, that introduce basic skills for arguing and proving. Here, it is important to emphasize the meaning of informal arguments in order to stress the quality of a given argumentation. If we highlight the possibility to formalize an informal argument we also underline the function and value of using algebra and variables in mathematics.

Using Anthropological Theory of Didactics (Chevallard 2003), Job and Schneider (2014) consider the development of calculus as an epistemological transition between two types of praxeologies: pragmatic and deductive. In this view, the authors discuss the dichotomy between formal and intuitive aspects of limits, as a mathematical activity that is expected to become rigorous on some formal definition. The authors highlight the notion of limit as a pragmatic model of magnitudes relying on mental objects. They argue that the prevalence of empirical positivism constitutes an obstacle to the learning of calculus, which is reinforced by the institutions, as a consequence of their inability to give credit to a pragmatic level of rationality.

2.4.4 Initiatives to Improve Teaching and Learning at the Beginning of the Tertiary Studies

Reported obstacles and pitfalls experienced by beginner undergraduate students in mathematics courses, as well as the acknowledgement of the first year of the university as a crucial step for long-term success in degree programs, led some institutions to implement initiatives aiming to ease the way into tertiary studies through the first year, to reduce retention and to increase graduation rates. Some of these initiatives were objects of research projects as we discussed in previous sections (e.g. Bausch et al. 2014; Duah et al. 2014).

Furthermore, Engelbrecht and Harding (2015) report a multi-dimensional approach conceptualized and implemented at the University of Pretoria, South Africa, aiming to improve teaching and learning in first year mathematics courses. This effort was motivated by the national policy of increasing graduates in the sciences in the country. The authors propose a framework, founded in research literature, in which five key aspects are identified and addressed towards the facilitation of students' transition during the first year at the university. These aspects regard mathematical preparedness, social transition, learning style, support, and conceptual understanding. The author suggests that problems associated with the transition to the university are perhaps exacerbated because of the procedural approach followed at school and the practice of examination coaching. Therefore, students have to undergo a change in thinking approach, from procedural to more conceptual, and a culture of independent learning needs to be fostered. Especially, in the context of mathematics teacher education programs several studies suggest innovative approaches at the entry phase aiming towards the improvement of professional knowledge by altering the conditions in university teaching (see for example, TEDS program, Buchholtz and Kaiser 2013).

2.5 Theoretical-Methodological Advances

The review we conducted for this survey leads to an observation that research in mathematics education at the tertiary level is a constantly growing field that have been endorsing a broad range of theoretical and methodological lenses including a significant shift of attention to sociocultural, institutional and discursive approaches (see, for example the Research in Mathematics Education special issue on "Institutional, sociocultural and discursive approaches to research in university mathematics education", Nardi et al. 2014a). In this section we will focus on some examples of theoretical and methodological advances we identified in our review, although by no means does this section intend to be a comprehensive summary of all the approaches we met.

2.5.1 Theoretical Perspectives and Liaison of Theories

Very recently, there has been increasing interest in discursive approaches into research, with a significant influence being made by the publication of Sfard's (2008) book on the Commognitive framework (see also Nardi et al. 2014b). Our review identified recent studies using the Commognitive framework for the analysis of university teachers' discursive practices in calculus courses (Park 2015; Viirman 2014, 2015); and of Calculus textbooks (Park 2016). Furthermore, Güçler (2016) suggests an "instruction that explicitly attends to the metalevel rules of the discourse on functions has the potential to support student learning" (p. 391). Also in the mathematical communication field, some recent studies focus on gestures and semiotics to describe university mathematics teachers' practices (Lee et al. 2009; Weinberg et al. 2015; Wheeler and Champion 2013).

When the *community* is the main focus, the lens of *communities of practice* has been used in studies that investigate teaching and learning practices at the tertiary level (see Biza et al. 2014). In this review we reported the study of Biza and Vande Hey (2014) that endorses this theoretical perspective. Especially from the community of inquiry perspective we visited Jaworski and Matthews (2011) who address teaching mathematics to engineering students.

Furthermore, the *documentational perspective* suggested by Gueudet and colleagues bring new theoretical constructs that can deal not only with the mathematics teacher actual instructional activity but with their activity outside the classroom in their preparation, evaluation and revision of their teaching resources as well as with their professional development (Gueudet 2015; Gueudet et al. 2014).

Additionally, *Anthropological Theory of Didactic* (*ATD*, Chevallard 2003; Winsløw et al. 2014) that analyze *institutionally* conditioned relationships in knowledge and practice and the *Theory of Didactic Situations* (TDS, Brousseau 1996; González-Martín et al. 2014) that consider the implicit or explicit rules of the interactions taking place within a system formed by the teacher, the students and the milieu, have been the frame of studies in tertiary mathematics education.

Finally, Rasmussen et al. (2015) are interested in the investigation of mathematical progress at both collective and individual level. To this aim they expand Cobb and Yackel's (1996) interpretive framework for coordinating social and individual perspectives and offer a set of four constructs for the examination of mathematical progress at a collective and individual level: "disciplinary practices, classroom mathematical practices, individual participation in mathematical activity, and mathematical conceptions that individuals bring to bear in their mathematical work" (p. 259). We return to the coordination of both individual and collective theoretical lenses later in this section.

In our review we identified, also, a range of studies, which aim to liaise different theoretical perspectives in order to address their research questions. We consider this as a step forward in the investigation of teaching and learning issues at university mathematics education in a more holistic way. Inevitably the combination of different theoretical perspectives is not always straightforward, as it demands *compatibility* between the epistemological and ontological underpinnings of distinct perspectives (Kidron 2016).

One example is coming from a recent study of González-Martín (2015) who investigates the use of textbooks by pre-university teachers in Quebec with particular focus on the concept of series of real numbers. To this aim the study deploys a combination of theoretical lenses by drawing on the Documentational approach (Gueudet et al. 2012, 2014) and the ATD (Chevallard 2003) in order to investigate how teachers interact with the textbooks and how the institutional environment in which they act establishes (and sometimes imposes) a set of conditions and constraints.

Another example is from the work of Tabach et al. (2015) who are interested in individual/collective interaction in a mathematical process in which four first year undergraduate STEM students work together towards the reinvention of the fundamental idea and technique of Euler's method. To this aim they draw on the *Abstraction in Context* (Hershkowitz et al. 2001) for the analysis of processes of constructing knowledge by individuals, and small groups and the *Documenting Collective Activity* (Rasmussen and Stephan 2008) for identifying normative ways of reasoning with groups of students. They highlight the need for methodologies that "provide a fine-grained analysis of the individual and collective processes that make inquiry learning possible" (Tabach et al. 2015, p. 2254).

2.5.2 Methodological Advances

In this section we draw on two quite different innovative methodological advances we met in our review: *storytelling* and *eye-movement* methodologies. In terms of storytelling, Nardi (2016) supports that "storytelling is an engaging way through which lived experience can be shared and reflected upon" and suggests a narrative approach of *re-storying* to present data (and analyses of these data) collected from interviews with twenty university mathematicians from six UK departments (Nardi 2008). Results are presented in the form of a dialogue between two fictional, yet entirely data-grounded, characters: a mathematician (M) and a researcher in mathematics education (RME). The re-storying approach is presented in details and exemplified through an application of it in a small number of interviews, which were re-storied into an exchange of utterances between the M and RME characters on potentialities and pitfalls of visualization in university mathematics teaching. Nardi (2016) concludes with a suggestion of "re-storying as a vehicle for community [M and RME community] rapprochement achieved through generating and sharing research findings".

There are an increasing number of studies in teaching and learning of mathematics adopting *eye-tracking* methodologies. Beitlich et al. (2014) used eye movement recording with eight mathematicians in order to identify how much attention adults with high expertise in mathematics pay to additional pictures when reading a written mathematical proof. They found that all participants paid attention to the pictures and tried to integrate information from text and picture by alternating between these representations. Chumachemko et al. (2014) compared eye movements in reading information in the Cartesian coordinate system of three groups of participants with different levels of mathematical competencies (mathematics graduates, non-mathematics undergraduates and secondary school students) and they found that the experts in comparison to novices have the ability of using additional essential information and to discard unnecessary data. Similarly, Obersteiner et al. (2014) recorded eye movements of eight mathematicians to identify their strategies on fraction comparison problems and they suggest this methodology for the investigation of individual strategies in problem solving. Finally, the Hodds et al. (2014) we discussed earlier, used eye movement methodology to assess how a booklet containing self explanation training, designed to attract students' attention on logical relationships within a mathematical proof, can significantly improve their proof comprehension.

3 Summary and Looking Ahead

In summarizing we would say, that although our focus was mainly studies published after 2014, the volume and the spread of publications we visited was impressive. The main points we identified in this review are as follows:

- The increasing interest in teaching practices at the tertiary level; the influence of teachers' perspectives, background, and research practices on their teaching, the role of resources and professional development in teaching; and a range of alternative approaches other than lecturing to teaching at the tertiary level.
- Problems in teaching mathematics to non-mathematics students and suggestions for enhancement; application of mathematics and modelling to non-mathematics disciplines; and how mathematical concepts are addressed in non-mathematics programs.

- Opportunities afforded by textbooks; how students use and read these textbooks; how they use online resources; and how they perceive or act in the assessment.
- Transition to the tertiary level in relation to the differences in the mathematical content; students' attitudes, practices and performances; the need for abstract mathematical concepts and formal mathematical thinking; and initiatives that address the challenges of such transition.
- Theoretical and methodological perspectives and liaison of theories that address issues in relation to the teaching and learning mathematics at the tertiary level

The spectrum of foci of the studies we reviewed indicates the increasing interest in the field as well as the prominent ongoing need for more robust research of issues pertaining mathematics teaching and learning at tertiary level. We will conclude this review with a discussion of potential ways forward for future research in this field of enquiry.

One example of an under-investigated area is the transition of mathematics graduates to postgraduate studies. In an older study Herzig (2002) discusses how doctoral mathematics students experience the research practice in their institution and the challenges they face in their enculturation to researcher mathematicians' practices. In a different context, Nardi (2015) addresses the transition of mathematics graduates to postgraduate programs in mathematics education. The study discusses an intervention into the practices of post-graduate teaching and supervision in the field of mathematics education that facilitates students in their shift on how to read, converse, write, and conduct research in the largely unfamiliar to them territory of mathematics education. It seems that more research is needed into the characteristics of the transition from undergraduate to post-graduate studies, especially when the epistemology changes, for example, from undergraduate studies in mathematics to postgraduate studies in statistics, engineering, mathematics education, or mathematics teaching (for those who want to become teachers).

Another area that has attracted attention recently but it is still in the process of taking a shape, is the field of university teachers' *knowledge*. What do we mean by *knowledge*? How does this *knowledge* develop? How does this *knowledge* reflect on practice? Potentially we may want to investigate practice and knowledge together, e.g., in discursive practices.

Teaching practice development is another potential area. As Jaworski et al. (2015) suggest key questions for further research are: "In what ways can teaching be characterized so that university teachers can gain relevant insights to teaching processes and develop teaching?" and "How can theories of teaching be employed to aid the design and development of teaching?" (p. 103). Additionally, teacher professional development is an emerging area of research that seeks and has the potential to generate further theoretical and methodological advances.

Another recently emerged area in tertiary mathematics education is related to the teaching of mathematics to non-mathematicians. This is related to the contextualization of mathematical ideas and the needs of non-mathematics specialists, as we mentioned earlier: What are the mathematical aspects that are necessary or helpful to understand specific context meanings and to solve applied problems or tasks? In what sense are context related mathematical problems different to the mathematical problems in a mathematics education context? What are the difficulties of specific groups of students? How can knowledge and competence developments be described and analyzed effectively and validly? Also, teaching mathematics to non-mathematics programs is related to how teachers see mathematics and mathematics teaching to non-mathematics specialists. What are their epistemological perspectives about mathematics? How do they perceive students needs? How do these perspectives influence their teaching?

The investigation of the questions above is open and a range of events and fora (such as ICME-TSG2, RUME, CERME-TWG14, INDRUM, etc.) and journals (especially IJRUME) related to the research in the teaching and learning mathematics at tertiary level are the arena in which we look forward to seeing the results of such investigations.

References

Albano, G., & Pierri, A. (2014). Mathematical competencies in a role-play activity. In P. Liljedahl, C. Nicol, S. Oesterle, & D. Allan (Eds.), *Proceedings of the Joint Meeting of PME38 and PME-NA36* (Vol. 2, pp. 17–24). Vancouver, Canada: PME.

Alpers, B. (2011). Studies on the mathematical expertise of mechanical engineers. *Journal of Mathematical Modelling and Application, 1*(3), 2–17.

Alpers, B. A., Demlova, M., Fant, C. H., Gustafsson, T., Lawson, D., Mustoe, L. et al. (2013). *A framework for mathematics curricula in engineering education. A report of the mathematics working group.* Brussels: European Society for Engineering Education (SEFI).

Ariza, A., Llinares, S., & Valls, J. (2015). Students' understanding of the function-derivative relationship when learning economic concepts. *Mathematics Education Research Journal, 27* (4), 615–635.

Artigue, M., Batanero, C., & Kent, P. (2007). Mathematics thinking and learning at post-secondary level. In F. K. Lester (Ed.), *Second handbook of research on mathematics teaching and learning: A project of the national council of teachers of mathematics* (pp. 1011–1049). Charlotte, NC: Information Age Publishing.

Balacheff, N., & Gaudin, N. (2010). Modelling students' conceptions: The case of function. *Research in Collegiate Mathematics Education, 16*, 207–234.

Barnett, M. D., Sonnert, G., & Sadler, P. M. (2014). Productive and ineffective efforts: how student effort in high school mathematics relates to college calculus success. *International Journal of Mathematical Education in Science and Technology, 45*(7), 996–1020.

Barquero, B., & Bosch, M. (2015). Didactic engineering as a research methodology: From fundamental situations to study and research paths. In A. Watson & O. Minoru (Eds.), *Task design in mathematics education an ICMI study 22* (pp. 249–272). Springer International Publishing.

Barquero, B., Bosch, M., & Gascón, J. (2008). Using research and study courses for teaching mathematical modelling at university level. In D. Pitta-Pantazi & G. Pilippou (Eds.), *Proceedings of CERME5* (pp. 2050–2059). Larnaca, Cyprus: University of Cyprus and ERME.

Bassili, J. N., & Joordens, S. (2008). Media player tool use, satisfaction with online lectures and examination performance. *Journal of Distance Education, 22*, 93–108.

Bausch, I., Biehler, R., Bruder, R., Fischer, P. R., Hochmuth, R., Koepf, W., et al. (2014). *Mathematische Vor-und Brückenkurse. Konzepte, Probleme und Perspektiven.* Wiesbaden: Springer.

Beitlich, J. T., Obersteiner, A., Moll, G., Mora Ruano, J. G., Pan, J., Reinhold, S., & Reiss, K. (2014). The role of pictures in reading mathematical proofs: An eye movement study. In P. Liljedahl, C. Nicol, S. Oesterle, & D. Allan (Eds.), *Proceedings of the joint meeting of PME38 and PME-NA36* (Vol. 2, pp. 121–128). Vancouver, Canada: PME.

Bergsten, C., Engelbrecht, J., & Kågesten, O. (2015). Conceptual or procedural mathematics for engineering students–views of two qualified engineers from two countries. *International Journal of Mathematical Education in Science and Technology, 46*(7), 979–990.

Biehler, R., Kortemeyer, J., & Schaper, N. (2015). Conceptualizing and studying students' processes of solving typical problems in introductory engineering courses requiring mathematical competences. In K. Krainer & N. Vondrová (Eds.), *Proceedings of CERME9* (pp. 2060–2066). Prague, Czech Republic: Charles University in Prague, Faculty of Education and ERME.

Bing, T. J. (2008). *An epistemic framing analysis of upper level physics students' use of mathematics.* Ph.D. thesis, University of Maryland. Retrieved from http://drum.lib.umd.edu/bitstream/1903/8528/1/umi-umd-5594.pdf.

Biza, I., Jaworski, B., & Hemmi, K. (2014). Communities in university mathematics. *Research in Mathematics Education, 16*(2), 161–176.

Biza, I., & Vande Hey, E. (2014). Improving statistical skills through students' participation in the development of resources. *International Journal of Mathematical Education in Science and Technology, 46*(2), 163–186.

Blömeke, S., Hsieh, F. J., Kaiser, G., & Schmidt, W. H. (2014). International perspectives on teacher knowledge, beliefs and opportunities to learn. *Teachers education and development study in mathematics (TEDS-M).* Dordrecht: Springer.

Blum, W., & Leiß, D. (2007). How do students and teachers deal with modelling problems? In C. Haines, P. Galbraith, W. Blum, & S. Khan (Eds.), *Mathematical modelling: Education, engineering, and economics* (pp. 222–231). Chichester: Horwood.

Bressoud, D., & Rasmussen, C. (2015). Seven characteristics of successful calculus programs. *Notices of the American Mathematical Society, 62*(2), 144–146.

Bressoud, D., Mesa, V., & Rasmussen, C. (Eds.). (2015). *Insights and recommendations from the MAA national study of college calculus.* Washington, DC: The Mathematical Association of America.

Brousseau, G. (1996). L'enseignant dans la théorie des situations didactiques. *Actes de la VIIIe école d'été de didactique des mathématiques*, 3–46. IREM de Clermont-Ferrand.

Buchholtz, N., & Kaiser, G. (2013). Improving mathematics teacher education in Germany: empirical results from a longitudinal evaluation of innovative programs. *International Journal of Science and Mathematics Education, 11*(4), 949–977.

Businskas, A. M. (2008). *Conversations about connections: How secondary mathematics teachers conceptualize and contend with mathematical connections.* (Unpublished doctoral dissertation). Burnaby, BC, Canada: Simon Fraser University.

Chevallard, Y. (2003). Approche anthropologique du rapport au savoir et didactique des mathématiques. In S. Maury & M. Caillot (Eds.), *Rapport au savoir et didactiques* (pp. 81–104). Paris: Faber.

Chumachemko, D., Shvarts, A., & Budanov, A. (2014). The development of the visual perception of the cartesian coordinate system: An eye tracking study. In P. Liljedahl, C. Nicol, S. Oesterle, & D. Allan (Eds.), *Proceedings of the joint meeting of PME38 and PME-NA36* (Vol. 2, pp. 313–320). Vancouver, Canada: PME.

Cobb, P., & Yackel, E. (1996). Constructivist, emergent, and sociocultural perspectives in the context of developmental research. *Educational Psychologist, 31*, 175–190.

Cofer, T. (2015). Mathematical explanatory strategies employed by prospective secondary teachers. *International Journal of Research in Undergraduate Mathematics Education, 1*(1), 63–90.

Coupland, M., Dunn, P. K., Galligan, L., Oates, G., & Trenholm, S. (2016, in press). Tertiary Mathematics Education. In K. Makar, S. Dole, J. Visnovska, M. Goos, A. Bennison & K. Fry (Eds.), *Research in mathematics education in Australasia: 2012–2015* (RiMEA). Rotterdam: Sense Publishers.

Croft, T., Duah, F., & Loch, B. (2013). 'I'm worried about the correctness': Undergraduate students as producers of screencasts of mathematical explanations for their peers—lecturer and student perceptions. *International Journal of Mathematical Education in Science and Technology, 44*(7), 1045–1055.

Czocher, J. A. (2014). Towards building a theory of mathematical modelling. In P. Liljedahl, C. Nicol, S. Oesterle, & D. Allan (Eds.), *Proceedings of the joint meeting of PME38 and PME-NA36* (Vol. 2, pp. 353–360). Vancouver, Canada: PME.

Duah, F., Croft, T., & Inglis, M. (2014). Can peer assisted learning be effective in undergraduate mathematics? *International Journal of Mathematical Education in Science and Technology, 45* (4), 552–565.

Dweck, C. S. (2007). *Mindset the new psychology of success: How we can learn to fulfill our potential*. New York: Ballantine Books.

Engelbrecht, J., & Harding, A. (2015). Interventions to Improve teaching and learning mathematics in first year courses. *International Journal of Mathematical Education in Science and Technology, 46*(7), 1046–1060.

Engelbrecht, J., Bergsten, C., & Kågesten, O. (2015). Conceptual and procedural approaches to mathematics in the engineering curriculum: views of qualified engineers from two countries. *International Journal of Mathematical Education in Science and Technology, 46*(7), 979–990.

Ellis, J. (2014). Preparing future professors: Highlighting the importance of graduate student professional development programs in calculus instruction. In P. Liljedahl, C. Nicol, S. Oesterle, & D. Allan (Eds.), *Proceedings of the joint meeting of PME38 and PME-NA36* (Vol. 3, pp. 9–16). Vancouver, Canada: PME.

Ellis, J., Hanson, K., Nuñez, G., & Rasmussen, C. (2015). Beyond plug and chug: An analysis of calculus I homework. *International Journal of Research in Undergraduate Mathematics Education, 1*(2), 268–287.

Frejd, P., & Bergsten, C. (2016). Mathematical modelling as a professional task. *Educational Studies in Mathematics, 91*, 11–35.

Fukawa-Connelly, T. P., & Newton, C. (2014). Analyzing the teaching of advanced mathematics courses via the enacted example space. *Educational Studies In Mathematics, 87*(3), 323–349.

González-Martín, A. S. (2015). The use of textbooks by pre-university teachers. An example with infinite series of real numbers. In K. Krainer & N. Vondrová (Eds.), *Proceedings CERME9* (pp. 2124–2130). Prague, Czech Republic: Charles University in Prague, Faculty of Education and ERME.

González-Martín, A., Bloch, I., Durand-Guerrier, V., & Maschietto, M. (2014). Didactic situations and didactical engineering in university mathematics: Cases from the study of calculus and proof. *Research in Mathematics Education, 16*(2), 117–134.

González-Martín, A. S., Nardi, E., & Biza, I. (2011). Conceptually-driven and visually-rich tasks in texts and teaching practice: The case of infinite series. *International Journal of Mathematical Education in Science and Technology, 42*(5), 565–589.

Goodchild, S., & Rønning, F. (2014). Teaching mathematics at higher education. In H. Silfverberg, T. Kärki, & M. S. Hannula (Eds.), *Proceedings of NORMA14* (pp. 396–400). Turku, Finalnd: Finnish Research Association for Subject Didactics.

Grenier-Boley, N. (2014). Some issues about the introduction of first concepts in linear algebra during tutorial sessions at the beginning of university. *Educational Studies in Mathematics, 87*(3), 439–461.

Güçler, B. (2016). Making implicit metalevel rules of the discourse on function explicit topics of reflection in the classroom to foster student learning. *Educational Studies in Mathematics, 91*, 375–393.

Gueudet, G. (2008). Investigating the secondary–tertiary transition. *Educational Studies in Mathematics, 67*(3), 237–254.

Gueudet, G. (2015). University teachers' resources and documentation work. In K. Krainer & N. Vondrová (Eds.), *Proceedings of CERME9* (pp. 2138–2144). Prague, Czech Republic: Charles University in Prague, Faculty of Education and ERME.

Gueudet, G., Buteau, C., Mesa, V., & Misfeldt, M. (2014). Instrumental and documentational approaches: From technology use to documentation systems in university mathematics education. *Research in Mathematics Education, 16*(2), 139–155.

Gueudet, G., Pepin, B., & Trouche, L. (Eds.). (2012). *From text to 'lived' resources: Mathematics curriculum materials and teacher development.* New York: Springer.

Hare, A., & Sinclair, N. (2015). Pointing in an undergraduate abstract algebra lecture: Interface between speaking and writing. In K. Beswick, T. Muir, T., & J. Wells (Eds.), *Proceedings of PME39* (Vol. 3, pp. 33–40). Hobart, Australia: PME.

Harris, D., Black, L., Hernandez-Martinez, P., Pepin, B., Williams, J., & with the TransMaths Team. (2015). Mathematics and its value for engineering students: What are the implications for teaching? *International Journal of Mathematical Education in Science and Technology, 46*(3), 321–336.

Hayward, C. N., Kogan, M., & Laursen, S. L. (2015, online first). Facilitating instructor adoption of inquiry-based learning in college mathematics. *International Journal of Research in Undergraduate Mathematics Education,* 1–24.

Herzig, A. H. (2002). Where have all the students gone? Participation of doctoral students in authentic mathematical activity as a necessary condition for persistence toward the PH.D. *Educational Studies in Mathematics, 50*, 177–212.

Hershkowitz, R., Schwarz, B., & Dreyfus, T. (2001). Abstraction in context: Epistemic actions. *Journal for Research in Mathematics Education, 32*, 195–222.

Hester, S., Buxner, S., Elfring, L., & Nagy, L. (2014). Integrating quantitative thinking into an introductory biology course improves students' mathematical reasoning in biological contexts. *CBE—Life Sciences Education, 13*, 54–64.

Heublein, U., Richter, J., Schmelzer, R., & Sommer, D. (2014). Die Entwicklung der Studienabbruchquoten an den deutschen Hochschulen. *Statistische Berechnungen auf der Basis des Absolventenjahrgangs 2012 (Forum Hochschule 4|2014).* Hannover: DZHW. Retrieved Feb 25, 2016 from http://www.dzhw.eu/publikation/forum.

Hieb, J. L., Lyle, K. B., Ralston, P. A., & Chariker, J. (2015). Predicting performance in a first engineering calculus course: Implications for interventions. *International Journal of Mathematical Education in Science and Technology, 46*(1), 40–55.

Hochmuth, R., Biehler, R., & Schreiber, S. (2014). Considering mathematical practices in engineering contexts focusing on signal analysis. In T. Fukawa-Connelly, G. Karakok, K. Keene & M. Zandieh. *Proceedings of RUME17* (pp. 693–699). Denver, Colorado.

Hochmuth, R., & Schreiber, S. (2015). Conceptualizing societal aspects of mathematics in signal analysis. In S. Mukhopadhyay & B. Greer (Eds.), *Proceedings of MES8* (pp. 610–622). Portland: Ooligan Press.

Hodds, M., Alcock, L., & Inglis, M. (2014). Self-explanation training improves proof comprehension. *Journal for Research in Mathematics Education, 45*, 62–101.

Iannone, P., & Simpson, A. (in press). University students' perceptions of summative assessment: The role of context. *Journal of Further and Higher Education*.

Iannone, P., & Simpson, A. (2015a). Students' preferences in undergraduate mathematics assessment. *Studies in Higher Education, 40*, 1046–1067.

Iannone, P., & Simpson, A. (2015b). Students' views of oral performance assessment in mathematics: Straddling 'assessment of' and 'assessment for' learning divide. *Assessment and Evaluation in Higher Education, 40*, 971–987.

Inglis, M., Palipana, A., Trenholm, S., & Ward, J. (2011). Individual differences in students' use of optional learning resources. *Journal of Computer Assisted learning, 27*(6), 490–502.

Jaworski, B., Mali, A., & Petropoulou, G. (2015). Approaches to teaching mathematics and their relation to students' mathematical meaning making. In K. Beswick, T. Muir & J. Wells (Eds.), *Proceedings of PME39* (Vol. 3, pp. 97–104). Hobart, Australia: PME.

Jaworski, B., & Matthews, J. (2011). Developing teaching of mathematics to first year engineering students. *Teaching Mathematics and its Applications, 30*, 178–185.

Jaworski, B., Robinson, C., Matthews, J., & Croft, A. C. (2012). An activity theory analysis of teaching goals versus student epistemological positions. *International Journal of Technology in Mathematics Education, 19*(4), 147–152.

Job, P., & Schneider, M. (2014). Empirical positivism, an epistemological obstacle in the learning of calculus. *ZDM—The International Journal on Mathematics Education, 46*(4), 635–646.

Jones, I., & Alcock, L. (2013). Peer assessment without assessment criteria. *Studies in Higher Education, 39*(10), 1774–1787.

Jukić Matić, L., & Dahl, B. (2014). Retention of differential and integral calculus: a case study of a university student in physical chemistry. *International Journal of Mathematical Education in Science and Technology, 45*(8), 1167–1187.

Kaiser, G., & Brand, S. (2015). Modelling competencies: Past development and further perspectives. In G. A. Stillman, W. Blum & M. Salett Biembengut (Eds.), *Mathematical modelling in education research and practice*, pp. 129–149. Springer International Publishing.

Kempen, L., & Biehler, R. (2014). The quality of argumentations of first-year pre-service teachers. In P. Liljedahl, C. Nicol, S. Oesterle, & D. Allan (Eds.), *Proceedings of the joint meeting of PME38 and PME-NA36* (Vol. 3, pp. 425–432). Vancouver, Canada: PME.

Kidron, I. (2016). Epistemology and networking theories. *Educational Studies in Mathematics, 91*(2), 149–163.

Kim, M. (2011). *Differences in beliefs and teaching practices between international and U.S. domestic mathematics teaching assistants*. Retrieved from ProQuest Dissertations and Theses. (885228899).

Larsen, S., Marrongelle, K., Bressoud, D., & Graham, K. (in press). Understanding the concepts of calculus: Frameworks and roadmaps emerging from educational research. In J. Cai (Ed.), *The compendium for research in mathematics education*. Reston VA: National Council of Teachers of Mathematics.

Le, A., Joordens, S., Chrysostomou, S., & Grinnell, R. (2010). Online lecture accessibility and its influence on performance in skills-based courses. *Computers and Education, 55*(1), 313–319.

Lee, H. S., Keene, K. A., Lee, J. T., Holstein, K., Early, M. E., & Eley, P. (2009). Pedagogical content moves in an inquiry-oriented differential equations class: Purposeful decisions to further mathematical discourse. *Proceedings of RUME12*. Raleigh, NC.

Leontiev, A. N. (1978). *Activity, consciousness and personality*. Englewood Cliffs: Prentice Hall.

Lithner, J. (2000). Mathematical reasoning in task solving. *Educational Studies in Mathematics, 41*, 165–190.

Lithner, J. (2003). Students' mathematical reasoning in university textbooks exercises. *Educational Studies in Mathematics, 59*, 29–55.

Lithner, J. (2004). Mathematical reasoning in calculus textbooks exercises. *Journal of Mathematical Behavior, 23*, 405–427.

Loch, B., & Lamborn, J. (2016). How to make mathematics relevant to first-year engineering students: Perceptions of students on student-produced resources. *International Journal of Mathematical Education in Science and Technology, 47*(1), 29–44.

Mali, A. (2015). Characterising university mathematics teaching. In K. Krainer & N. Vondrová (Eds.), *Proceedings of CERME9* (pp. 2187–2193). Prague, Czech Republic: Charles University in Prague, Faculty of Education and ERME.

Mali, A., Biza, I., & Jaworski, B. (2014). Characteristics of university mathematics teaching: Use of generic examples in tutoring. In P. Liljedahl, C. Nicol, S. Oesterle, & D. Allan (Eds.), *Proceedings of the joint meeting of PME38 and PME-NA36* (Vol. 4, pp. 161–168). Vancouver, Canada: PME.

Mesa, V. (2010). Strategies for controlling the work in mathematics textbooks for introductory calculus. *Research in Collegiate Mathematics Education, 16*, 235–265.

Mills, M. (2015). Business faculty perceptions of the calculus content needed for business courses. *Proceedings of RUME18* (pp. 231–237). Pittsburgh, Pennsylvania.

Mkhatshwa, T. P., & Doerr, H. M. (2015). Students' understanding of marginal change in the context of cost, revenue, and profit. In K. Krainer & N. Vondrová (Eds.), *Proceedings of CERME9* (pp. 2201–2206). Prague, Czech Republic: Charles University in Prague, Faculty of Education and ERME.

Mustoe, L. (2002). Mathematics in engineering education. *European Journal or Engineering Education, 27*(3), 237–240.

Nardi, E. (2008). *Amongst mathematicians: Teaching and learning mathematics at university level*. New York: Springer.

Nardi, E. (2016). Where form and substance meet: using the narrative approach of re-storying to generate research findings and community rapprochement in (university) mathematics education. *Educational Studies in Mathematics, 92*(3), 361–377.

Nardi, E. (2015). "Not like a big gap, something we could handle": Facilitating shifts in paradigm in the supervision of mathematics graduates upon entry into mathematics education. *International Journal of Research in Undergraduate Mathematics Education, 1*(1), 135–156.

Nardi, E., Biza, I., González-Martín, A. S., Gueudet, G., & Winsløw, C. (2014a). Institutional, sociocultural and discursive approaches to research in university mathematics education. *Research in Mathematics Education, 16*(2), 91–94.

Nardi, E., Ryve, A., Stadler, E., & Viirman, O. (2014b). Commognitive analyses of the learning and teaching of mathematics at university level: the case of discursive shifts in the study of Calculus. *Research in Mathematics Education, 16*(2), 182–198.

Niss, M. (2003). Mathematical competencies and the learning of mathematics: The Danish KOM project. In A. Gagatsis & S. Papastravidis (Eds.), *3rd mediterranean conference on mathematics education* (pp. 115–124). Athens, Greece: Hellenic Mathematical Society and Cyprus Mathematical Society.

NRC. (2003). *BIO2010: Transforming undergraduate education for future research biologists*. Washington, DC: National Academies Press.

NRC. (2009). *A new biology for the 21st century: Ensuring the United States leads the coming biology revolution*. Washington, DC: National Academies Press.

Obersteiner, A., Moll, G., Beitlich, J. T., Cui, C., Schmidt, M., Khmelivska, T., & Reiss, K. (2014). Expert mathematicians' strategies for comparing the numerical values of fractions—evidence from eye movements. In P. Liljedahl, C. Nicol, S. Oesterle, & D. Allan (Eds.), *Proceedings of the joint meeting of PME38 and PME-NA36* (Vol. 4, pp. 337–344). Vancouver, Canada: PME.

Oh Nam, K. (2015). How to teach without teaching: An inquiry-oriented approach in tertiary education. In K. Beswick, T. Muir, T., & J. Wells (Eds.), *Proceedings of PME39* (Vol. 1, pp. 19–36). Hobart, Australia: PME.

Park, J. (2015). Is the derivative to function? If so, how do we teach it? *Educational Studies in Mathematics, 89*(2), 233–250.

Park, J. (2016). *Communicational approach to study textbook discourse on the derivative, 91*, 395–421.

Petropoulou, G., Potari, D., & Zachariades, T. (2011). Challenging the mathematician's 'ultimate substantiator' role in a low lecture innovation. In B. Ubuz (Ed.), *Proceedings of PME35* (Vol. 3, pp. 385–392). Ankara, Turkey: PME.

Petropoulou, G., Jaworski, B., Potari, D., & Zachariades, T. (2015). How do research mathematicians teach Calculus? In K. Krainer & N. Vondrová (Eds.), *Proceedings of CERME9* (pp. 2221–2227). Prague, Czech Republic: Charles University in Prague, Faculty of Education and ERME.

Polya, G. (1949). *How to solve it: A new aspect of mathematical method*. Princeton, New Jersey: Princeton University Press.

Raman, M. (2002). Coordinating informal and formal aspects of mathematics: Student behavior and textbook messages. *Journal of Mathematical Behavior, 21*, 135–150.

Raman, M. (2004). Epistemological messages conveyed by three high-school and college mathematics textbooks. *Journal of Mathematical Behavior, 23*, 389–404.

Ramful, A., & Narod, F. B. (2014). Proportional reasoning in the learning of chemistry: Levels of complexity. *Mathematics Education Research Journal, 26*(1), 25–46.

Rasmussen, C., Ellis, J., Zazkis, D., & Bressoud, D. (2014). Features of successful calculus programs at five doctoral degree granting institutions. In P. Liljedahl, C. Nicol, S. Oesterle, & D. Allan (Eds.), *Proceedings of the joint meeting of PME38 and PME-NA36* (Vol. 5, pp. 33–40). Vancouver, Canada: PME.

Rasmussen, C., & Stephan, M. (2008). A methodology for documenting collective activity. In A. E. Kelly, R. A. Lesh, & J. Y. Baek (Eds.), *Handbook of design research methods in education: Innovations in science, technology, engineering, and mathematics learning and teaching* (pp. 195–215). New York: Routledge.

Rasmussen, C., & Wawro, M. (in press). Post-calculus research in undergraduate mathematics education. In J. Cai (Ed.), *The compendium for research in mathematics education*. Reston VA: National Council of Teachers of Mathematics.

Rasmussen, C., Wawro, M., & Zandieh, M. (2015). Examining personal and collective level mathematical progress. *Educational Studies in Mathematics, 88*(2), 259–281.

Rønning, F. (2016). Innovative education in mathematics for engineers: Some ideas, possibilities and challenges. In R. Göller, R. Biehler, R. Hochmuth, & H.-G. Rück (Eds.), *Didactics of mathematics in higher education as a scientific discipline—conference proceedings*. Kassel: Universitätsbibliothek Kassel.

Serrano, L., Bosch, M., & Gascón, J. (2010). Fitting models to data. The mathematising step in the modelling process. In V. Durand-Guerrier, S. Soury-Lavergne & F. Arzarello (Eds.), *Proceedings of CERME6* (pp. 2185–2196). Lyon: INRP 2010.

Sfard, A. (2008). *Thinking as communicating. Human development, the growth of discourse, and mathematizing*. New York, NY: Cambridge University Press.

Shepherd, M., Selden, A., & Selden, J. (2012). University students' reading of their first-year mathematics textbooks. *Mathematical Thinking and Learning, 14*(3), 226–256.

Shepherd, M. D., & van de Sande, C. C. (2014). Reading mathematics for understanding—from novice to expert. *Journal of Mathematical Behavior, 35*, 74–86.

Singletary, L. M. (2012). *Mathematical connections made in practice: An examination of teachers' beliefs and practices*. (Unpublished dissertation). Athens, GA: University of Georgia.

Smith, G., Wood, L., Coupland, M., Stephenson, B., Crawford, K., & Ball, G. (1996). Constructing mathematical examinations to assess a range of knowledge and skills. *International Journal of Mathematical Education in Science and Technology, 27*(1), 65–77.

Solomon, Y., Croft, T., Duah, F., & Lawson, D. (2014). Reshaping understandings of teaching-learning relationships in undergraduate mathematics: An activity theory analysis of the role and impact of student internships. *Learning, Culture and Social Interaction, 3*(4), 323–333.

Speer, N., Smith, J., & Horvath, A. (2010). Collegiate mathematics teaching: An unexamined practice. *The Journal of Mathematical Behavior, 29*(2), 99–114.

Suominen, A. L. (2015). *Abstract algebra and secondary school mathematics: Identifying mathematical connections in textbooks.* (Unpublished dissertation). Athens, GA: University of Georgia.

Tabach, M., Rasmussen, C., Hershkowitz, R., & Dreyfus, T. (2015). First steps in re-inventing Euler's method: A case for coordinating methodologies. In K. Krainer & N. Vondrová (Eds.), *Proceedings of CERME9* (pp. 2249–2255). Prague, Czech Republic: Charles University in Prague, Faculty of Education and ERME.

Tall, D. (1992). The transition to advanced mathematical thinking: Functions, limits, infinity and proof. In D. A. Grouws (Ed.), *Handbook of research on mathematics teaching and learning* (pp. 495–511). New York: Macmillan.

Tallman, M., Carlson, M., Bressoud, D., & Pearson, M. (in press). A characterization of Calculus I final exams in U.S. colleges and universities. *International Journal of Research in Undergraduate Mathematics Education.*

Treffert-Thomas, S. (2015). Conceptualising a university teaching practice in an activity theory perspective. *Nordic Studies in Mathematics Education, 20*(2), 53–77.

Tolley, H., & MacKenzie, H. (2015). Senior management perspectives on mathematics and statistics support in higher education, sigma report: http://www.sigma-network.ac.uk/wpcontent/uploads/2015/05/sector-needs-analysis-report.pdf.

Tuminaro, J., & Redish, E. F. (2007). Elements of a cognitive model of physics problem solving: Epistemic games. *Physical Review Special Topics—Physics Education Research, 3*, 1–21.

Ufer, S. (2015). The role of study motives and learning activities for success in first semester mathematics studies. In P. Liljedahl, C. Nicol, S. Oesterle, & D. Allan (Eds.), *Proceedings of the joint meeting of PME38 and PME-NA36* (Vol. 4, pp. 265–272). Vancouver, Canada: PME.

Uhden, O., Karam, R., Pietrocola, M., & Pospiech, G. (2012). Modelling mathematical reasoning in physics education. *Science and Education, 21*(4), 485–506.

Ulriksen, L., Madsen, L. M., & Holmegaard, H. T. (2010). What do we know about explanations for drop out/opt out among young people form STM higher education programmes? *Studies in Science Education, 46*(2), 209–244.

Usiskin, Z. (1974). Some corresponding properties of real numbers and implications for teaching. *Educational Studies in Mathematics, 5*, 279–290.

Viirman, O. (2014). The functions of function discourse: University mathematics teaching from a commognitive standpoint. *International Journal of Mathematical Education in Science and Technology, 45*(4), 512–527.

Viirman, O. (2015). Explanation, motivation and question posing routines in university mathematics teachers' pedagogical discourse: A commognitive analysis. *International Journal of Mathematical Education in Science and Technology, 46*(8), 1165–1181.

Weber, K. (2015). Effective proof reading strategies for comprehending mathematical proofs. *International Journal of Research in Undergraduate Mathematics Education, 1*(3), 289–314.

Weber, K., & Mejia-Ramos, J.-P. (2011). Why and how mathematicians read proofs: An exploratory study. *Educational Studies in Mathematics, 76*, 329–344.

Weinberg, A., Fukawa-Connelly, T., & Wiesner, E. (2015). Characterizing instructor gestures in a lecture in a proof-based mathematics class. *Educational Studies in Mathematics, 90*(3), 233–258.

Weinberg, A., & Wiesner, E. (2011). Understanding mathematics textbooks through reader-oriented theory. *Educational Studies in Mathematics, 76*(1), 49–63.

Weinberg, A., Wiesner, E., Benesh, B., & Boester, T. (2012). Undergraduate students' self-reported use of mathematics textbooks. *Problems, Resources, and Issues in Mathematics Undergraduate Studies, 22*(2), 152–175.

Wenger, E. (1998). *Communities of practice: Learning, meaning and identity.* Cambridge: Cambridge University Press.

Wheeler, A., & Champion, J. (2013). Students' proofs of one-to-one and onto properties in introductory abstract algebra. *International Journal of Mathematical Education in Science and Technology, 44*(8), 1107–1116.

White, N., & Mesa, V. (2014). Describing cognitive orientation of Calculus I tasks across different types of coursework. *ZDM—The International Journal on Mathematics Education, 46*(4), 675–690.

Wilkie, K., & Tan, H. (2015). Performance or progress? Influences on senior secondary students' mathematics subject selection. In P. Liljedahl, C. Nicol, S. Oesterle, & D. Allan (Eds.), *Proceedings of the joint meeting of PME38 and PME-NA36* (Vol. 4, pp. 305–312). Vancouver, Canada: PME.

Winsløw, C., Barquero, B., De Vleeschouwer, M., & Hardy, N. (2014). An institutional approach to university mathematics education: From dual vector spaces to questioning the world. *Research in Mathematics Education, 16*(2), 95–111.

Yang, K.-L. (2012). Structures of cognitive and metacognitive reading strategy use for reading comprehension of geometry proof. *Educational Studies in Mathematics, 80*, 307–326.